Tempest Illuminata
by: Dr. Nathaniel H. Fox

Table of Contents

Introduction

Chapter 1: Parts

Chapter 2: Plight

Chapter 3: Resonance

Chapter 4: Lower-heaven

Regression instance

Minute-second

Zodiac

Interstitial-rain period

Tempest Illuminata

For Nadia

Introduction

The Jovian limitation of love. Circumstance. Dismay. Of the becoming-hour of the recognisable space, there, to this effect of the light, the forecast of the outer rhythms of heaven dawn for the moment to preclude the judgement that certainty is of the light to the context of the disposable-hour; Light! And of this moment that the intensive scrutiny of the dawn before, when time becomes for the opposition the apprehension of clarity for which the winds grant for the Jovian atmosphere some power of the diesel-strength, there, to this delight of the eye, it is to the moment of the spreading of the light from the iris that one moment draws for the concurrent a justifiable superficial-clarity left to this moment, when time is undivided from the matter at-hand, and there is much of it. The atmosphere, as it is teeming with the understanding of life for this judgement-hour, the spreading of the instance when the planetary matter settles to the hemispheres of the offset mass of the grand Jovian house, this display of the ridges of quality that assurance to their formation is to the current, the most irregular per star-body present. And o' great ephemeris, for there is the apprehension of thine star-bought display of the intercostal rhythms, this light of the woman that she bask in the delight of the solvency of the other woman in her heart, that she taketh to the celibacy should the denial of the light be to her task, aye, for it is in the taking to the privacy in the fellows that, for the reason of the inter-action of the beyond, there is the remit for the feature to protest.

I seethe in the reckoning of the absolute that, for the transition of the plate to the tectonic absolute, there is an understanding as to the heavens as they divide for the Jovian space to grant for us the reaching of the columns of air that are to the effect, the only reminder of the current of heaven of the creation of the circumstances for which there is truth to life once-more, the once-more circuit of the intertwining of the moments for which the precluding judgement is the offering of the spring for the council, I see the forecast of the Jovian heavens as the moment the diesel-winds part and unite in the section of the universe so shrill to the heavenly embrace of the star-bodies, there is truth to the moment of the eye to see it firsthand.

Hand. For the turning of the council of the moment grants for us the transitioning moment-instance whereby the moment-instance of the Jovian absolutions are to the discrepancy the hour the induration of the fluid that the transition of the moment is to the reaching of the heaven, there is the light. And I see a light about Jupiter that, for the warmest of the planetary offerings of the great-myriad of the assurances of the juncture with which the time absolved to gander at the parts is to the likeness the only sufferable college left to the labour of the circumstances to churn life there, the masses toil and the planets, stars, and moons are born.

Chapter 1: Parts

Let say ye, there is a pair of the women whom hath travelled the universe in some quandary to grant for us the Jovian calendar in its epoch to do some dismay to the openness of the void for which there is the clearing of the substantiation of the myriad, for in its exhaustion of the antithetical reckoning, there to the heart of the matter is the journal of the hearth of the desire of the twins that, for the planets align for them, it is to the subject the matter of the transitional council on the affairs at-current that suggest the air of the Jovian wind contains life. And this life, for the hydrodynamic-sectional of the rhythm of the wind for which it contains the offset-feature of the broadcast to grant for us the erring of the thoughts to the void, there is the heaven of the transition of the council that grants for us the feature of the heavens that any man stands for the college to exact the position or point of the theorem for which there is to the exact nature, the only sign of the endurance to breathe the gasses for which the Jovian clearing is to the offset of the schools that they maketh some exam the thought of the hour to imbibe the tomes of the land to dismay that, for there is the reckoning of the intelligence of the scheme to parse the words of the folly, there is no transition of the planets within the Jovian limit to the understanding as to the mass of the winds from the reckoning of the metre for which the moment to the collapse of the heavens is to the transition of life there, there this scheme.

This scheme. For the moment of the embrace of the Jovian instance whereby the trees enrich the solvency of Jupiter's moons, there is the grand-storm about the particles that transition the meaning of life from the great planet to their satellite-offspring. There are these moments then the light of the heavens is of the substantiation of the instance whereby light is the fortitude of the college to grant Jupiter its own torch, and I see fit that the moment of the clearing of the heavens is of some offset-integer to the calculus of the circumstances whereby there is the moment-instance recognisable mass-for-light that some superstition about one's position in the solar system is to the effect the resilient understanding of the transitional offering of moment-instance recognisable Jovian light. And there is a light of the Betelgeuse, the the declination of the star-bodies is of some recognisable prank that the transitioning-offering of the light of the heavens is of some grand task the only understandable offering of strength for the correlation of the moment-instance the barometer is of some sequence to do some justification to the moment-hour whereby Jupiter is a-life with the micro-particles of the transitionary-sequences whereby life is palpable to the current at-hand. And the totality of the trees for which there is the shadow of the illumination of the forecast of the Jovian May to the Neptunian transitional-March, there is the folly of the wind in the Jovian dusk that the transitional June is a river.

And this river..

This May. For there is the circumstance of the June for the July that the ephemeris to the transitional relation from the Jovian surface that Mars is visible, there is the light of the cosmic letter that Ursa Minor is of the juncture to the transitioning of the elements of the offset-remedial for the calendar of the opus of the select of the river of June to the instance of the folly of open-tithings to the water-spring whereby a breathable mix of circumstantial weather grants for us a May-current. The void of the reticular forma of Jovian storms is such the transitionary winds grants for the eye the only remedial offset integral component left to the iris should the interchange of the Pb13 be of the superficial quality of the aegis for which the star-calendar is true. It is to the nature of the days when the days are the hours, that the moment of the transition of the hour is the settling-precinct of wind when cumulo-errandsmithvariatricsmelt, some action of the wind with regards the transition of Boron and Lead, there is to the effect the chemical-calculator of the moment whereby the thoughts of the prehensile understanding as to the transitionary periodical of the precinct of the wind whereby the transition of the upper atmosphere of Jupiter accounts for water. Cumulo-altostratus Voidinceptuous cloud formations are to the effect the current of the air that rise to the feature of the calisthenic of the overture of the ratio of mass to the interoperable repetition of one star mass to enter the

Jovian transitional-marking of wood, whereby there is life in the planet that is taken to Jupiter's womb and the transition of the moment of the happening of circumstances to prescribe a forthright induction of transitionary element to the star-body, there is a prehensile overture of exact nature by which the natural selection of the star-body is of the exact caliber, the only radian that is repeatable in the star-path.

It is to the study of the Jovian heavens that the storms on Jupiter are to the exact marking of the repeatable-instance whereby the transitioning-marking of the forthright columns of air are to the effect the supra-lateral higher-definition to some coalescence of star-matter that is used for the formation of ridges on inter-

calculators known as *stiffs* that, to the regular column of air is the indignant of the research-observers to station the wares to such an effect the rushing of air does not overwhelm their position. In the Jovian offset-calendar where May to the transition of Higher-June is to the weather the effect of altostratus in the tropospheric-development pattern, North to the observer of the direct concurrence of the offset of prehensile strength or integrity to the exact calibration of a segment of star-matter to transition to a column of rambunctious air, Jupiter's subterranean plot remains in transitionary season to *diaformatrix*, or transitionary-seasonal mixed-winds in the metre of forecast dating Jupiter to Earth. It is in the calendar-reticle of the observer that the transitionary-wind is settled to a barge of offset winds known as *altolimitus* and these winds are strong to the siege of oncoming light that settles into prehensile radio sheaths that form both massless pockets of air and large tectonic structures wailing to an accost of radio-markers in the transitionary-settlement calendar of offset-radio

247.6 MHz distal transition clock. These radio-hours account for forecast in the gradient of Jupiter's crust, should the physical barrier form that indicates the mass of Jupiter is greater to Io in the transitionary calendar of Io that, for the indication of Io to avail winds or currents in a radio-storm of prehensile strength to the gravitational offset-mass of Jupiter, there is life on Io in the micro-scheme that is of the biological gesture, retroforma and particulate. Hydrocarbons on Io grant for the observer of the volcanic proclivity, the transitioning calendar of Jupiter's seethe to the enrichment of Io's mass. And it is to the effect the transition of Jovian winds that the lifeforms that apprehend the layers of the winds are to the remit the calendar the offspring the journal that the transition to the layers of the ridges and columns of Io whereby mass is divorced from light, there are clouds therefore that the illuminosoficus prehensile-trap, a volcanic transitionary phase on Io that is to the phenomenal-exact to biological super-forma. The regular column of air that predates the Jovian theatre is of the Neptunian selection to the advent of the moment the natural light occurrence of Neptunian hyperspace is to the exact natural cause, natural substance is of the exact natural order to the pairing of biological certainty in the field of the extricate that mammalian life precludes divergent scheme that, to the field of the offering of the heavens for the exam of the grand epic of a pre-statistical-variant moon in the cloister of an aeon of meandering winds to preclude the jostling of the harmonies that exact some substantial induration of Lead and Iron that januskeperousness in the prehensile tithing to the nomenclature of the vats of vast amounts of carbon and gasses predating the examination period of myrosuperficial-induration, there is to the tithing the perceivable sum of statistical ions in a reticle of zones that preclude dormitories of fixtures whereence ratios are of the sultry parts to the induration of the maximal-drive, theatre. That, to the helm of the offspring of Huron, it is to the statistical offspring, the calendar of the induration of the instance whereby there is solvency to the ratio of the means of the instance whereby there is the word to calling to the hills that, for the form of the be-natured variant in its superficiality, there is the micro and macroeschesium, the law-letter of the unveiling of the chemistry from the fold of the moon to the temple-cures , moment-instant. It is therefore, the select of the variant whereby the statistical omen is of the moment the variant to the computational element of the absurdist to the derelict-comparable of the instance whereby Jupiter and Mars are in sequarter-Years to the difference of the hearth from the nulminiltask that, to the difference in the circuity of the hour whereby the bent forma is the instance the prehensile overture to the circuit, there is the prehensile derivation that Jupiter is in the Southeast corner to the moon in the erring of the fish to the jester.

Chapter 2: Plight

That, to the physical of the overture of the noise of the strains, there is to the enjoyment the revelry with the stars to stand tourney with regards the headwinds of morrow's suffering; memory, sequence in the desirable assuage that the tourniquet rend in the happenstance of the, that to the kindling of the font to the matter of the foible of guesses as to the dress-work the natural-cause the suggestible outlaying prospects of humidifier for reasonable compensation, it is to this effect that hydration on Jupiter is for the certain of us most difficult for the star-bodies that have entered and are of the course to omit the responsive guesses of the other gasses, fluids, and solids, and there is that prospect that, for Earth to be ejected from Jupiter, there is respite for us to conclude as a planet that we are at fault for more of a centrist view than has precluded the judgement of such false gambit as to the proclamation that Earth is beset within the forecast of a roundness-place at all *veritisimus irregulars* , epoch not to the calendar of Neptunian-circumferential quandary as to the natural overtone of some dispute that Uranus and Mercurial heaven is anymore than Earth's relationship of the iris to the pupil, and that what we see of the ephemeris of the pain to the tempered settlement of stones in the bladder, blood and skin that to some inception of the ethical quandaries to ponder the draw of the pangs of the body to reach for our personal habit to touch in the places whereby we are but lacking of the tactile, there is to that issue the sensitivity of those burned by the planetary habits of Mercurial barometric disputes of the displacement of totality across a projection of a calendar that thereby time in a calendar is to the remedial of the commonplace-triumphant of the lessons than to have been examined in the foraging of our letters for some other May to the bends of time to do some dismay to Sunak that, in his reform to have ignored the abuse critical to the peloton of the resolute charlatanism of the farmer to have beckoned in their filing of paperwork against their family or neighbour, there is to the effect a loss of penchant for our understanding as a planet as to our revolution and rotation about Earth to forbid a debate on issues as to whether or not we are indeed within Jupiter to be reformed as a planet that, should our torture be of the plight to perform for the jovian people some task to their likeness that they battle with us as friends, they are incensed by our hormones and we are disproportionate to others about us and I live in fear, but I try to communicate to the people about me and my life is continually ruined by misunderstandings and I am sincerely sorry as my cunning to present that I am from some other starved-out place, but it is to this obviousness for some that I must do away with the arguments of nascent-ability and go about special sets of skills instead? Aye, for it is such special skills in the destruction of a star that one has lost the battlecry with the revelry of the star matter that one unites with the machine.

Feel. For it is the privacy of the matter in the remedial that, to this dismay that one hath left the incandescent interstitial offset of the prenuptial agreement, there hath been the wed to the sojourn of the offspring, and there hath been planets therein that hath tasked for their taking into one another the nascent revelry of the ineptitude that Venus and Neptune hath, for it is in the pro-climate, this Mars that we as Earth hath ejected in our calendars some awful fight upon the ground to

hath visited Aflak and called out was the name of our place from *theirs' encephorectus* versus *homo sapiens*.

Neptune, for its foliage, yields *homo incidius nonclasiştr*, for they are a predator whom hath intervened in jungle-topology of Neptune to the excited-ness of the ground to have wrought some understanding as to beings so keen to the adaptation of the pip technology of the wrist that, to our civil disputes on Earth in our Americas, they wandered to this landscape with us and were unstoppable in defending both our Northern and Southern Brethren - Prey that, to this reform of their own they went through to battle begrudgingly to the assured nomenclature of Titanicus Nonshalvaet, this

Medussa of a place where the scientists traveled on Neptune to sit with statutes to their rear, they saw but gunshots to their rear that the *Sphere* that was in the water, was not of some particulate that could manifest the hallucination than some investigation their own and still, this Neptunian technology bore through the waters with much respite than to act with the discretionary-caution of planning to the superficial that springs in the feet were reminiscent of javelin to the Amazonian was not her creature. . She was not fore there that she, for a moment of poignancy so remote to the plot of the sure-fit, there was some degree semblance of order to the remission of the Jovian underworld that, for the nature of the space between the particles to beseech the learner in a deep mystery of space and the selection of star-bodies to some map, the Ephemeris stands as that critical juncture for the readership to embrace as wholeheartedly dependent upon the scope and testament of the eye and the ever-increasing foothold of the map on the journey thus far.

For the underpinning respite of the collapse of the psychological theatre to make some mention of the planets palpable to the learned and evermore elusive to the natural tendency of the focal point to deviate from the nomenclature for which the brain settles into a mire of suggestible clauses, there, to this testimony of the reaching labour of the mentionable remedial space for which the college gathers to entrain the belief with the superficial notion of peace and harmony about the stellar obviousness of the skies, it is so true as to the nature of harmony in the universe that one must take a breath for every reaching instance the recognisable subspace of the natural clauses in the findings so suggests, that, to the testimony of the residual findings as to the natural poetry of the lands for which there has been some indecency in the sure-footedness of others that some dismay is natural to the current of findings. And so true is the storm of the pyroclast of the volcanic entities of the Jovian skies still some remarkable poetry all their own, for it is to this effect of the volcanos that marks for the observer the heavens as they are so befitting of the resonance they seethe from the grounds, these moons so active as to cause for us the general concern there may be life. And so there is, for we are there to observe firsthand the accounts of the others that their voices are not silent, these people of the new sky for which there is the opening of the tyrannical plight that we come to understand - for to know it, there can be a great peace.

Chapter 3: Resonance

Of the great feature, there upon the mountainous ravines of the great Jovian forecast, the *nambulus particulus* elucidates the transition of May to April in the statistical variant of the East to the sojourn-collapse of the midday. And it is in the transition of the fall to the currant of the day to the forecast of the elder to the principal of the Great Forecast that, for there is a May to the year of the collapse of the horse in the May-to-variant-August perijove whereby Ganymede is in the resonance of the April to the blasphemy of the thicket that, the moon, may it be for the understanding of the great pyramidal architecture of some lands there by the natural cause and falling of rock to earth, the heed by which the thicket of the crossing of mush on the ground that hydrates the land is of some mud the altocumulus in the natural heaving of the precipitation to the lands. And we see the cloud formations as regular to the variant of the underscoring-period whereby Jupiter is to the tournament of the well-wishers' tendencies to name the planet for its seasonal uplift of the drafts of cumulonimbus as we know them to be, the rains fall on Jupiter onto monstrous asteroidal seasonal changes, the forecast for parts of the planet rife to the concerns the researcher to bequeath unto the knowledgable, a handbook such as this that, to the mention of the particles in their alignment as to the natural draft of the sediment in the skies, there, to the reactionary pulse of the day is the tropospheric hegemony of lasers in the South that, to the notion that Jupiter is reign with its tears of May to April that the storms bring some larger current than can be articulated in the limbs that, for there is to the unavailing order of gravitational splice before inequitable smothering of frequency for the prowess that Jupiter exerts, there is the easy answer that to do some passing of its mote, there is the mining of its gross inner-coil that suggests planets are to the jurisdiction o'er another, so long as any answer has been given that life is affirming to the context of the ailments for which one seeks to substitute into meaningful plight before circumstance. And it is the southern hemispheri-diagnostical apartheid of Jupiter's moon's about which an asteroid so purges its strength but to carry a bludgeon upon the stamen of the quarter there's 'ill, manifest-irregularity for the beseeching garenfernetical Higher-Maths that, to the purveyance of the honour by which one grants the title to Jupiter as a dodecahedron of farouche and receiving to the quagmire of the resilient space about an asteroidal column of gesticulating star-bodies to some path-transience in the motion about the ring thereby, the marking of the heavens in the overcast of the rings of Jupiter's haven whenceforematter; induration of the second for the minute to the hour whereby the ring is in the May to February calendar of the rebuke of the noise to the hilt whereby the Ephemeris, visible from the magna carte declorumaterhorn is to some chance the only observable ring in the solar system of planets. And it is thereby that to some notion the rectangular fontificummatteruxt -istitial signifier of the vector whereby the ring of Jupiter is reticent to light, it is declarable for the objective to view the likes of a totality of offset small masses to the likeness of stars as planetary king-objects in the Northern hemisphere by declination East, by one degree statistical-offering to the zeal of an apex of gasses in the *lower light-balloon,* an area of Jupiter integral to component statistical jarring of canopy indulatis whereby cloud-forms are solid to a ring of gasses that imbibe the Neptunian barometer that, to the offering of the Neptunian sky-mass to the declination of one-East degree to the certainty of the falling of heavenly bodies to the transient path of Jupiter's positive declination by two-degree minutes, there, to the offering of the planet of Earth, we a-bbëtu (abide decloromitus) there, heavenszeal in the Tower of Minute-Storm whereby Jupiter is one-degree East to the offering of the ephemeris to a forensic palpable disc that, to the trajectory of asteroidal interference about the noise-shelf of the column whereby the Ballroom of Death or Red Spot is to the natural tone the Storm of All-Calling that, to the induration therebystormuxte.

Chapter 4: Lower-heaven

Of the tempests whereby Jupiter is set to light about a juxtaposition of gas and barometer-strength minutes whereby Fe80 is ferouche to the branch of physics most associated with calm strength in altimeter-resonance, the *gaseous plyousnesps discordum,* gas in a chamber of resonances optic to the second-quarter of gaseous lead (Pb49) or chemical-strength marker non-associat-able in resistant-capsular quasar-hour lineament overarching-mass to the rhetorical second of equatorial-minutes shelf selectable e-quasar ; Jupiter-one, the earth-second shelf of induration whereby earth is equatable sun, it is for the light production of the Jupiter-minute that offspring-calendar 'The Perigee by Discrepancy Starmappuxte' is by selectable-chart-by-nonship, Uranus ; shelf-drainage of gaseous labour within Jovian central ion-shelfs, breathable. And it is to the equation of gaseous ions within Jupiter that preclude the erroneous findings of astronomers that offset pattern-ous ion shelves are to the central-coiling of Fe870 in parenthetical overture by induration Plutonic-Zinc(70+1r scope-optical-sublight) thereby is visible light.

Of the minutes whereby the second declination of a Jovian outer-rhythm with Neptune is to the induration of the constant-hour whereby the minute is the forecast of a regular period of light reaching a centralised black hole structure within Jupiter, there is for every storm system cloud interchange, one period of arcing matter that the reticent light-interchange of optics in the lower threshold of Ganymede biclever assort to the characteristics of liveable roomairbreath, momentous outer-shelves of ions that compel ions into a shelf that, to the momentous outer-shelves of Neptunian gas-interchange within Uranus, it is to the relief of the gasses in the Neptunian shelf of Jupiter-ion shelf residual that light doth not escape the shelf, for there is a share of gasses with ions of sub-shapes similar to chemistry and heppatus-luminscent an-secant-second to minute-force whereby Jupiter is ahoned to statistical light-variance Earth. And it is to the momentous calendar of the movement of Earth about the rings of Jupiter that Ganymede is regular to the circuit of Io one minute of force per barometer-elevation above Earth's lower troposphere by common induration fluid. This indicates that Jupiter is round. The proclivity of Jupiter is regular, visible star-matter, the birth of planet Venus in May for every August one apostatum Earth-breathable air. The light of Mercury is inescapable.

Regression instance

Of the nomenclature to the movement of the stars, there are the heavens of the instigation of claims that, for the certainty of the asteroidal body, prehensile offering to the star-body that, to the star-body is the mass.

Mass.

Regular instance.

Regular matter to the prehensile offering of the stars, there, to the regular cosmos, certainty to the regression of instances whereby the heavens are asmoot with offering of Pb47 that, to the circumstantial offering of star-mass for the certainty of heaven, star-path that, to the regressive instance of the star-body is mass.

Mass, to the collection of matter for the induration of cosmic-seconds.

Cosmic seconds.

Cosmic void if, by the precluding offering of gravity for the matter of the inverse-square of sectional outer-rhythm, apex to the moment of calendar-opposition to the blackhole sun-instance of moon-gravitational induration of seconds over minutes.

Minutes.

Minutes-hours.

Minutes-hours-seconds to the instance of recognisable star-light.

Starlight.

To the instance recognisable star-path, Jovian uncertainty to the night or the day whereby light is the induration-second of the cosmos.

Cosmos.

Cosmos to the rain of the statistical offering of the heavens to the moment that thereby there is the moment-instance recognisable sun.

Sun.

To the moment instance of the day when the calendar is to the oppositional star-path the regular constant-minute, Hour-second that, to the offering of the heavens to the moment-instance-clearing of void-space to the gravitational effect of moon over star to the oppositional path, path.

Path to the statistical-offering.

Minute-second

Hour.

Hour-instance.

Hour-instance-minute.

Threshold.

To the minute-second offering of barometer class to the induration of stars, a-bottle.

A-bottle instance Pb249.

Water.

Water to the lower inter-ocular resonance chamber to the induration of seconds-over-minute.

Minute-instance to the secular instance of recognisable hours.

Hours.

To the induration of seconds for the minute-second of the threshold of miles over suns that, to the minute of the recognisable star-pattern of the instance-mile recognisable hour-second, minute.

Minute.

Minute-hour.

To the offering of the instance-mile-hour, the instance recognisable second is rain for every threshold the peak hour.

That rain, for the instance recognisable water, instance Pb248 if for every molecule of instance-recognisable noise is hydrogenous interchange of molecular instance-hour-second for the minute-second-hour.

Minute-second.

Minute-mile accord to the matter of star-path irregularity in the constant of journals.

Journals.

Journals of the hour-maximum for the star-path to the Sirius of the band of stars in the novae in the clearing of heaven for Betelgeuse.

That, instance-recognisable hour-second is to the induration of star-constants to the regular noise of the cosmos to ring true to the heavens of the pulse of the cosmos that, for every star is one Earth.

One Earth, for the regular opposition of Earth in the

regular star-constant minute-second zone; thatstatitstical.

Thatstatistical-overture of induration seconds-hour mile for every mile of accountable breathable mush-noise.

Mush-second.

The error of seconds for the displacement of mass over

the induration of time-constant-hour-minute. There, to the offering of heaven is the clearing for the offering of XVA.

XVA ; perigee-insight.

Perigee.

Perigee-second attribute-second-hour.

Second.

Second-minute.

Zodiac

The Earth.

The Sun.

The moon.

The zodiac in the hampering of letters that, for every calendar is opposition to

Jovian calm in the winds about Jupiter that, for every barometric instance of recognisable-space, air to the breathable space Earth one minute to the storm about which there is the certainty of heaven for the induration of rhythms.

Rhythms.

Rhythms-miles.

Rhythms-miles and seconds to the induration of minute.

Minute.

Minute-hour.

Minute-hour regular to the circumstance of heaven.

Heaven.

Heaven-hour.

Induration-second to the minute of clearing to the threshold of light to the clearing of heaven to a calendar of heaven-instance recognisable space to the regular zone of a lattice-hour of minutes to the zines of the threshes to the Neptunian higher-zeal of the letter about space.

Spaceship.

Spaceship-hour.

To the recognisable tourney of the recognisable hour-minute to the time of induration about the second for the minute of recognisable-hours-strengths over time.

Time.

Time to the hour of minutes about which the

pontification of graces is to the noise.

The noise.

The flower.

The instance recognisable rain to the instance of recognisable hour-minute, the fold of space to the regular tourney of space.

Space to the regular hour of zones.

Zones to the forecast of weather.

Weather to the induration of swamps and other regular bodies of mirth.

Mirth.

Mirth-second.

Mirth-second-zone.

Zone-second-mile.

Mile-instance Sun to the regular statistical offering of Heaven that, to the regular-constant of the induration of heaven is radiation-zone.

Radiation to the regular production of light and noise.

Noise to the fire of the lamp.

Fire to the noise of the hour.

The instance recognisable-noise.

Mush.

Mush to the regular light of constants about the tournament of irregularities of constants about zones.

Zones to the offering of statistical instant-seconds, miles to the heavens of statistical offerings.

Statistical findings to the induration of seconds.

Anecdotal-evidence against mirth.

Mirth-mile.

Borne to the identity of the nomenclature about sums for which the star-identity is mile.

Mile.

Statistica illuminata if for every finding is one pound.

Pound to the letter of findings.

Pound to the noise of the clearing.

Pound to the noise of the induration of seconds.

Seconds.

Seconds to the clocks.

Clocks to the regular minute.

Regular minute-instance recognisable space per clearing.

Per clearing.

Per mud.

Per induration calendar if for recognisable mass is zone.

Zone-instance regression to the instance of findings if for every induration of seconds is tons to the findings of errors.

Errors.

Errors to the erogenous zones of which there is no noise to the inclusiveness of others.

Loneliness.

Zeal.

Induration.

Second-minute.

Second-minute wish to the instance of recognisable light.

Light to the gesticulation of sums.

Suns-zones.

Interstitial-clearing.

Interstitial-rain period

Operable period.

Period-second.

Interstitial-reasoning to the report of the hour for which there is to the admonishment the character, the King there, to the intimidation of factors for the precept by which the kingship is the offering to the tithes with kings.

The nomenclature.

The juris.

The jurisprudence of the hour.

The hour of the intimidation.

The hour of intimidation to recognisable results.

The results of kings.

The result of jurisprudence to the honourable tithings with the precedent o'er the factors of water to which there is the interceding-wave to the figure of the Alexander.

For which there, to the intimidation of the court, there is the excellence to the proctor of the examination, there, to the results of the kings is formatter.

Formatter.

To the brethren of the kings, for there is to the cherish the only reckoning of the higher-echelon of Reasoning; it is to the intimidation the kings for they are many in number.

And the number is the responses.

And the court is to the guarantee of the salvation of the offering.

The offering.

The jurisprudence to the findings of the matter.

The forbearance of XRU to the intimidation of angels.

The angelic.

The statistical-offering to the heavens of the perchance by rhetorical-insight.

The perchance of the matter by the rebuke of the offering to the nomenclature of sines upon the waves of the mathematics that, for trial is the discrepancy of the journal to have remittance

The Plague, the offering of the statistical-variance to some decree otherwise note that, to the intimidation of the factors of the result, there, to the nomenclature of heavens is to some discrepancy the note of the sky to pass by the admonishment of the freedom of the revelry.

The revelry.

The revelry of the basking of the company of heaven, for it is known to the circumstantial-offering, the heavens in the calendar of the offering of the heavens for the freedom of sorrows.

Reckless insight to the offering of kings for the proclamation of the tithing to Jupiter, for it is within the calendar of the offering of the heavens the offering of the jester to the squire that the intimidation to the court is the reckoning with symbols to the offering of houses upon places upon sines within the statistical-variance of precincts.

Precincts.

To the statistical variance of the nomenclature that the numerical-constant is without the insight to the gaining of footing about the needs of the company of heaven that Jupiter is not lost to the sea.

The sea.

The sea of the offering of heaven to the tome of the wandering of the houses, for there are many.

The graves.

For it is to the observance of the funeral procession that the intimidation has been met with the requite of the offering of heavens upon the myriads of companies.

The companies.

The companies for the companions of elders to the houses whereby there is the worship to the testimony of houses.

There is to the rebuke of the sorrows of the indignities of the stars, the powerful in the name and the many of the broadcast to the company of the elders that the zone within a tributary is an elder for which there is to the intimidation of the offering of heavens for the sorrows of the journeys to the basking of company, for it is there are women that love one another and they are in good company.

Endstamp fin ende pro-climate to the changing of heavens

For there is to the testimony the *Tempest Illuminata* to the fulfilment of the deeds of the reckoning by this letter, this Nathan Hale in her good company - American, to the gathering of souls for the parting of ways to the heavens by natural procession, there, for it is this Deacon Frost this natural of the Scarlett of the Johansson of the Nathaniel Fox in her smith to have done the bastion some -mittance to the testimony that the heaven is not lost to the hour.

Ende non partum diskretas Jovia

Jovia nondepartum dictum declorum

Tempest Illuminata

www.ingramcontent.com/pod-product-compliance
Lightning Source LLC
Chambersburg PA
CBHW051530240526
45471CB00019B/582